Grove Karl Gilbert

The History of the Niagara River

Grove Karl Gilbert

The History of the Niagara River

ISBN/EAN: 9783743332423

Manufactured in Europe, USA, Canada, Australia, Japa

Cover: Foto ©berggeist007 / pixelio.de

Manufactured and distributed by brebook publishing software
(www.brebook.com)

Grove Karl Gilbert

The History of the Niagara River

THE HISTORY

OF

THE NIAGARA RIVER,

By G. K. GILBERT.

EXTRACTED FROM THE SIXTH ANNUAL REPORT OF THE COMMISSIONERS OF THE STATE RESERVATION AT NIAGARA, FOR THE YEAR 1889.

ALBANY:
JAMES B. LYON, PRINTER.
1890.

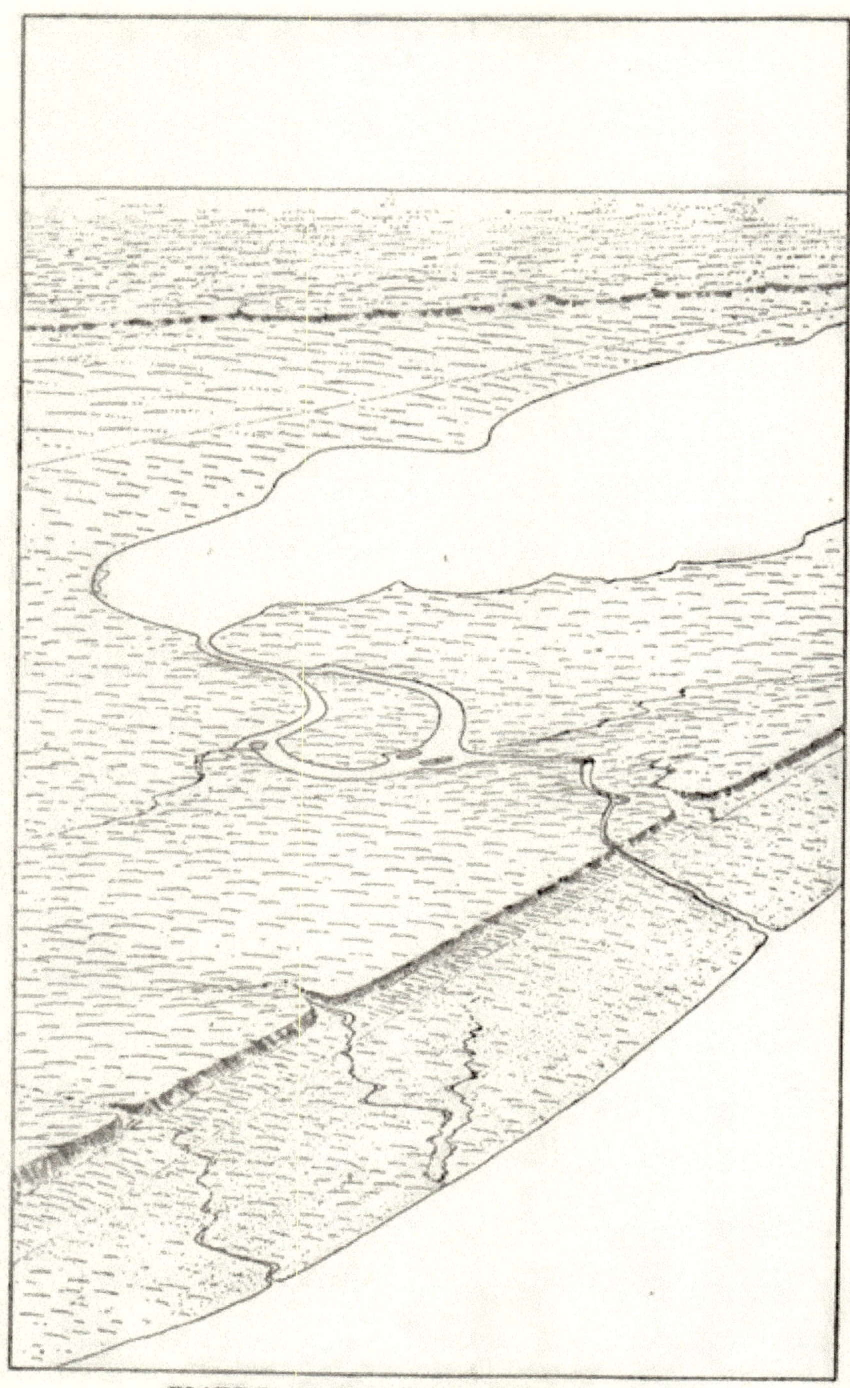

PLATE I.—Bird's-eye view of Niagara river.

THE HISTORY OF THE NIAGARA RIVER.[1]

By G. K. Gilbert.

The Niagara river flows from Lake Erie to Lake Ontario. The shore of Erie is more than 300 feet higher than the shore of Ontario; but if you pass from the higher shore to the lower, you do not descend at a uniform rate. Starting from Lake Erie and going northward, you travel upon a plain — not level but with only gentle undulations — until you approach the shore of Lake Ontario, and then suddenly you find yourself on the brink of a high bluff or cliff overlooking the lower lake, and separated from it only by a narrow strip of sloping plain. The birds-eye view in Plate I is constructed to show the relations of these various features, the two lakes, the broad plateau lying a little higher than the shore of Lake Erie, the cliff, which geologists call the Niagara Escarpment, and the narrow plain at its foot.

Where the Niagara river leaves Lake Erie at Buffalo and enters the plain, a low ridge of rock crosses its path, and in traversing this its water is troubled; but it soon becomes smooth, spreads out broadly, and indolently loiters on the plain. For three-fourths of the distance it can not be said to have a valley, it rests upon the surface of the plateau; but then its habit suddenly changes. By the short rapid at Goat Island and by the cataract itself the water of the river is dropped 200 feet down into the plain, and thence to the cliff at Lewiston it races headlong through a deep and narrow gorge. From Lewiston to Lake Ontario there are no rapids. The river is again broad, and its channel is scored so deeply in the littoral plain that the current is relatively slow, and the level of its water surface varies but slightly from that of the lake.

The narrow gorge that contains the river from the Falls to Lewiston is a most peculiar and noteworthy feature. Its width

---

[1] This essay contains the substance of a lecture read to the American Association for the Advancement of Science at its Toronto meeting, August, 1889.

rarely equals the fourth of a mile, and its depth to the bottom of the river ranges from 200 to 500 feet. Its walls are so steep that opportunities for climbing up and down them are rare, and in these walls one may see the geologic structure of the plateau. They are constituted of bedded rocks — limestone, shale and sandstone — lying nearly horizontal, and a little examination shows that the same strata occur in the same order on both sides. So evenly are they matched, and so uniform is the general width of the gorge, that one might suspect, after a hasty examination, the two sides had been cleft asunder by some Plutonic agency. But those who have made a study of the subject have reached a different and better conclusion — the conclusion that the trench was excavated by running water, so that the strata of the two sides are alike because they are parts of continuous sheets, from each of which a narrow strip has here been cut.

The contour of the cataract is subject to change. From time to time blocks of rock break away, falling into the pool below, and new shapes are then given to the brink over which the water leaps. Many such falls of rock have taken place since the white man occupied the banks of the river, and the breaking away of a very large section is still a recent event. By such observation we are assured that the extent of the gorge is increasing at its end, that it is growing longer, and that the cataract is the cause of its extension.

This determination is the first element in the history of the river. A change is in progress before our eyes. The river's history, like human history, is being enacted, and from that which occurs we can draw inferences concerning what has occurred and what will occur. We can look forward to the time when the gorge now traversing the fourth part of the width of the plateau will completely divide it, so that the Niagara will drain Lake Erie to the bottom. We can look back to the time when there was no gorge, but when the water flowed on the top of the plain to its edge, and the Falls of Niagara were at Lewistown.

We may think of the river as laboring at a task — the task of sawing in two the plateau. The task is partly accomplished. When it is done the river will assume some other task. Before it was begun what did the river do?

How can we answer this question? The surplus water discharged from Lake Erie can not have flowed by this course to Lake Ontario without sawing at the plateau. Before it began the cutting of the gorge it did not flow along this line. It may have flowed somewhere else, but if so it did not constitute the Niagara river. The commencement of the cutting of the Niagara gorge is the beginning of the history of the Niagara river. We have accomplished somewhat of our purpose if we have discovered that our river had a beginning.

We are so accustomed to think of streams, and especially large streams, as permanent, as flowing on forever, that the discovery of a definite beginning to the life of a great river like the Niagara is important and impressive. But that discovery does not stand alone. Indeed, it is but one of a large class of similar facts familiar to students of geology. Let us consider for a moment the tendency of stream histories and the tendency of lake histories. Wherever streams fall over rocky ledges in rapids or in cataracts, their power of erosion is greatly increased by the rapid descent, and they deepen their channels. If this process continues long enough, the result must be that each stream will degrade its channel through the hard ledges until the descent is no more rapid there than in other parts of its course. It follows that a stream with cascades and water-falls and numerous rapids is laboring at an unfinished task. It is either a young stream, or else nature has recently put obstructions in its path.

Again, consider what occurs where a lake interrupts the course of a stream. The lower part of the stream, the outflowing part, by deepening its channel continually tends to drain the lake. The upper course, the inflowing stream, brings mud and sand with it and deposits them in the still water of the lake, thus tending to fill its basin. Thus, by a double process, the streams are laboring to extinguish the lakes that lie in their way, and given sufficient time, they will accomplish this. A stream whose course is interrupted by lakes is either a young stream, or else nature has recently put obstructions in its path.

Now if you will study a large map of North America, you will find that the region of the Great Lakes is likewise a region of small lakes. A multitude of lakes, lakelets, ponds and swamps where ponds once were, characterize the surface from the Great Lakes northward to the Arctic ocean, and for a distance southward into the United States. In the same region water-falls abound, and many streams consist of mere alternations of rapids and pools. Farther south, in the region beyond the Ohio river, lakes and cataracts are rare. The majority of the streams flow from source to mouth with regulated course, their waters descending at first somewhat steeply, and gradually becoming more nearly level as they proceed. At the south the whole drainage system is mature; at the north it is immature. At the south it is old; at the north, young.

The explanation of this lies in a great geologic event of somewhat recent date — the event known as the age of ice. Previous to the ice age our streams may have been as tame and orderly as those of the Southern States, and we have no evidence that there were lakes in this region. During the ice age the region of the Great Lakes was somewhat in the condition of Greenland. It was covered by an immense sheet of ice and the ice was in motion. In general it moved from north to south. It carried with it whatever lay loose upon the surface. It did more than this, for just as the soft water of a stream, by dragging sand and pebbles over the bottom, wears its channel deeper, so the plastic ice, holding grains of sand and even large stones in its under surface, dragged these across the underlying rock, and in this way not only scoured and scratched it, but even wore it away.

In yet other ways the moving ice mass was analagous to a river. Its motion was perpetual, and its form changed little, but that which moved was continually renewed. As a river is supplied by rain, so the glacier was supplied by snow falling upon regions far to the north. To a certain extent the glacier discharged to the ocean like a river, breaking up into icebergs and floating away; but its chief discharge was upon the land, through melting. The climate at its southern margin was relatively warm, and into this warm climate the sheet of ice steadily pushed and was as steadily dissolved.

Whatever stones and earth were picked up or torn up by the ice, moved with it to its southern margin and fell to the ground as the ice melted. If the position of the ice margin had been perfectly uniform, its continuously deposited load might have built a single high wall; but as the seasons were cold or warm, wet or dry, the ice margin advanced and retreated with endless variation, and this led to the deposition of irregular congeries of hills, constituting what is known as the "drift deposit." Eventually the warm climate of the south prevailed over the invader born of a cold climate, compelling it to retreat. The motion of the ice current was not reversed, but the front of the glacier was melted more rapidly than it could be renewed, and thus its area was gradually restricted. During the whole period of retrenchment, the deposition of drift proceeded at the margin of the ice, so that the entire area that it formerly occupied is now diversified by irregular sheets and heapings of earth and stone.

The ancient configuration of the country was more or less modified by the erosive action of the ice, and it was further modified by the deposits of drift. The destructive and constructive agencies together gave to the land an entirely new system of hills and valleys. When the ice was gone, the rain that fell on the land could no longer follow the old lines of drainage. Some of the old valleys had perhaps been obliterated; others had been changed so that their descent was in a different direction; and all were obstructed here and there by the heaps of drift. The waters were held upon the surface in innumerable lakes, each overflowing at the lowest side of its basin, and thus giving birth to a stream that descended to some other lake. Often the new lines of descent — the new water courses — crossed regions that before had had no streams, and then they were compelled to dig their own channels. Thus it was that the whole water system of a vast region was refashioned, and thus it has come to pass that the streams of this region are young.

Like every other stream of the district of the Great Lakes, the Niagara was born during the melting of the ice, and so we may begin our chronicle with the very beginning of the river.

If you will again call to mind the features of a general map of the United States and Canada, and consider the direction in which the streams flow, you will perceive that there is a continuous upland, a sort of main divide, separating the basin of the Great Lakes from the basin of the Mississippi. (A part of its course appears as a broken line on the maps in Plates IV and V.) It is not a mountain range. In great part it is a region of hills. In places it is only the highest part of the plain. But it is nevertheless a continuous upland, else the waters would not be parted along its course. When the ice had its greatest extent, it passed over this upland, so that the waters produced by its melting fell into the Ohio and other tributaries of the Mississippi, as well as into streams that discharged to Delaware and Chesapeake bays. Afterward, when the glacier gradually fell back, there came a time when the ice front lay in the main to the north of the great water-parting, but had not yet receded from the Adirondack mountains, so that the water that flowed from the melting glacier could not escape by way of the St. Lawrence river, but gathered as a lake between the upland divide and the ice front. In fact it formed not one but many lakes, each discharging across the divide by some low pass; and as the great retreat progressed, these lakes were varied in number and extent, so that their full history is exceedingly complex.

The surfaces of these lakes were stirred by the winds, and waves beat upon their shores. In places they washed out the soft drift and carved cliffs; elsewhere they fashioned spits and bars. These cliffs and spits and other monuments of wave work survive to the present time, and have made it possible to trace out and map certain of the ancient lakes. The work of surveying them is barely begun, but from what is known we may add a chapter to the history of our river.

There was a time when one of these lakes occupied the western portion of the basin of Lake Erie, and discharged across the divide at the point where the city of Fort Wayne now stands, running into the Wabash river and thence into the Ohio. The channel of this discharge is so well preserved that its meaning can not be mistaken, and the associated shore-lines have been traced for many miles eastward into the Ohio, and northward into Michigan. Afterward, this lake found some other point of discharge, and a new shore-line was made twenty-five feet lower. Twice again the point of discharge was shifted, and other shore-lines were formed. The last and lowest of the series has been traced eastward across the States of Ohio and Pennsylvania and into western New York, where it fades away in the vicinity of the town of Careyville. At each of the stages represented by these four shore-lines, the site of the Niagara was either buried beneath the ice or else submerged under the lake bordering the ice. There was no river.

The next change in the history of the lakes was a great one. The ice, which had previously occupied nearly the whole of the Ontario basin, so far withdrew as to enable the accumulated water to flow out by way of the Mohawk valley. The level of discharge was thus suddenly lowered 550 feet, and a large district previously submerged became dry land. Then for the first time Lake Erie and Lake Ontario were separated, and then for the first time the Niagara river carried the surplus water of Lake Erie to Lake Ontario.

The waves of the new-born Lake Ontario at once began to carve about its margin a record of its existence. That record is wonderfully clear, and the special training of the geologist has not been necessary to the recognition of its import. The earliest books of travel in western New York describe the Ridge road, and tell us that the ridge of sand and gravel which it follows was even then recognized by all residents as an ancient beach of the lake.[1] In the Province of Ontario the beach was examined and described by the great English geologist, Charles Lyell, during his celebrated journey in America,[2] and it afterward received more careful study by Mr. Sandford Fleming,[3] and by the geologists of the Canadian Survey.[4] In western New York it was traced out by the great American geologist, James Hall, during his survey of the geology of the fourth district of the State.[5] Within a few years more attention has been given to detail. Professor J. W. Spencer has traced the line continuously from the head of the lake at Hamilton past Toronto, Windsor and Grafton to the vicinity of Belleville,[6]

1 C. Schultz, Jr.: Travels on an inland voyage * * * in the years 1807 and 1808, New York, 1810, p. 85.

De Witt Clinton: Discourse before the New York Historical Society, 1811, p. 58.

Francis Hall: Travels in Canada and the United States in 1816 and 1817, Boston, 1818, p. 119.

2 Travels in North America in the years 1841-2. New York, 1845. Vol. 2, pp. 86-87.

3 Sandford Fleming: Notes on the Davenport gravel drift. Canadian Journal, New Series, vol. 6, pp. 247-253.

4 Geological Survey of Canada, report to 1863, pp. 914-915.

5 Natural History of New York. Geology, Part IV, pp. 348-354.

6 Communicated to the Philosophical Society of Washington: to be published in Vol. 11 of the Bulletin of the Society.

beyond which point it is hard to follow. South of the lake, I myself have traced it from Hamilton to Queenston and Lewiston; thence to Rochester, and all about the eastern end of the basin to Watertown, beyond which point it is again difficult to trace. Southeast of the present margin of Lake Ontario, there was a great bay, extending as far south as Cayuga lake, and including the basin of Oneida lake, and it was from this bay that the discharge took place, the precise point of overflow being the present site of the city of Rome. For this predecessor of Lake Ontario Professor Spencer has proposed the name of Iroquois.

Putting together the results of his survey and of my own, I have been able to prepare a map (Pl. II) exhibiting with a fair amount of detail the outline of the old lake. It will be observed that the northeastern portion of the shore is not traced out. In fact it is not traceable. The water was contained on that side by the margin of the glacier, and with the final melting of the ice all record of its shore vanished.

The form and extent of Lake Iroquois, and the form and extent of each other lake that bordered the ice front, were determined partly by the position of the pass over which the discharge took place and by the contour of the land; but they were also determined to a great extent by the peculiar attitude of the land.

Perhaps a word of general explanation is necessary in speaking of the attitude of the land. Geologists are prone to talk of elevation and subsidence—of the uprising of the earth's crust at one place or at one time, and of its down-sinking at another place or another time. Their language usually seems to imply the rise or fall of an area all together, without any relative displacement of its parts; but you will readily see that, unless a rising or sinking tract is torn asunder from its surroundings, there must be all about it a belt in which the surface assumes an inclined position, or, in other words, where the attitude of the land is changed. If the district whose attitude changes is a lake basin, the change of attitude will cause a change in the position of the line marked about the slopes of the basin by the water margin, and it may even cause the overflow of the basin to take a new direction.

The Ontario basin has been subjected to a very notable change of attitude, and the effect of this change has been to throw the

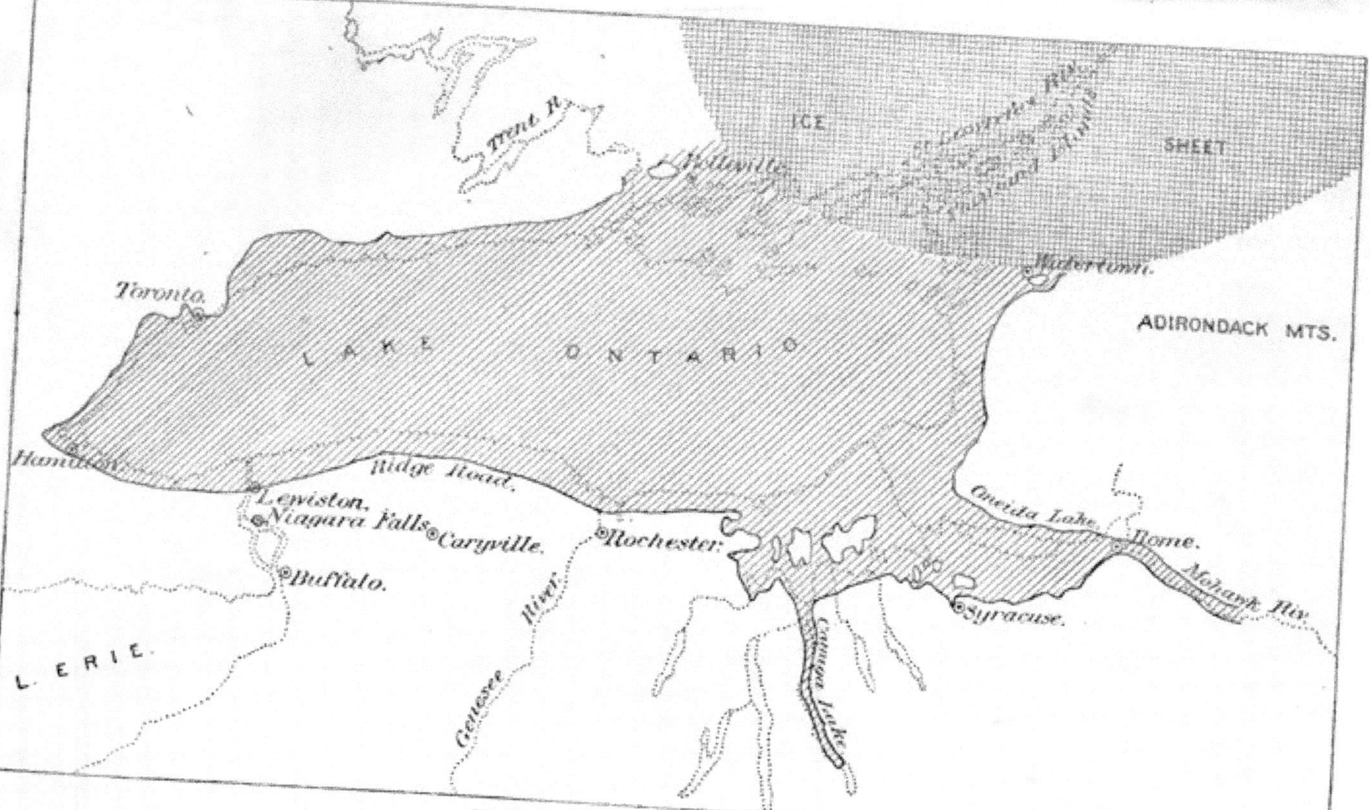

PLATE II.—Map of Lake Iroquois.

EXPLANATION.—Modern hydrography in dotted lines Ancient lake area shaded. Ice sheet cross-shaded.

ancient shore-line out of level. When the shore-line was wrought by the waves, all parts of it must have lain in the same horizontal plane, and had there been no change in the attitude of the basin, every point of the shore-line would now be found at the level of the old outlet at Rome. Instead of this, we find that the old gravel spit near Toronto — the Davenport ridge — is forty feet higher than the contemporaneous gravel spit on which Lewiston is built; at Belleville, Ontario, the old shore is 200 feet higher than at Rochester, N. Y.; at Watertown 300 feet higher than at Syracuse; and the lowest point, in Hamilton, at the head of the lake, is 325 feet lower than the highest point near Watertown. From these and other measurements we learn that the Ontario basin with its new attitude inclines more to the south and west than with the old attitude.

The point of discharge remained at Rome as long as the ice was crowded high against the northern side of the Adirondack mountains, but eventually there came a time when the water escaped eastward between the ice and the mountain slope. The line of the St. Lawrence was not at once opened, so that the subsidence was only partial. The water was held for short times at various intermediate levels, recorded at the east in a series of faint shore-lines. Owing to the attitude of the land, these shores are not traceable all about the basin, but pass beneath the present water level at various points.

Finally the ice blockade was raised in the St. Lawrence valley, and the present outlet was established. During the period of final retreat the attitude of the land had slowly changed, so that it was not then so greatly depressed at the north as before; but it had not yet acquired its present position, and for a time Lake Ontario was smaller than now, its western margin lying lower down on the slope of the basin.

An attempt has been made in Pl. III to exhibit diagramatically the relations of ice dams and basin attitudes to one another and to the river. The various elements are projected, with exaggeration of heights, on a vertical plane running a little west of south, or parallel to the direction of greatest inclination of old water-planes. At N is represented the Niagara escarpment and the associated slope of the lake basin; at A the Adirondack mountains. R and

T are the passes at Rome and at the Thousand Islands. Successive positions of the ice front are marked at I^1, I^2 and I^3. The straight line numbered 1 represents the level of lake water previous to the origin of the Niagara river; 2 gives the first position of the water level after the establishment of the Rome outlet; and the level gradually shifted to 3; 4 is the first of the series of temporary water levels when the water escaped between the mountain slope and the ice front; 5 represents the first position of the water level after the occupation of the Thousand Island outlet; and 6, the present level of Lake Ontario.

It should be added parenthetically that the shore of Lake Iroquois as mapped in Pl. II is not quite synchronous. Between 2 and 3 of Pl. III there was a continuous series of water levels, but it was not easy to map any one except the highest. The northern part of the map delineates the margin of water level 2, and the southern part the margin of water level 3.

It is easy to see that these various changes contribute to modify the history of the Niagara river. In the beginning, when the cataract was at Lewiston, the margin of Lake Ontario, instead of being seven miles away as now, was only one or two miles distant, and the level of its water was about seventy-five feet higher than at present. The outlet of the lake was at Rome, and while it there continued, there was a progressive change in the attitude of the land, causing the lake to rise at the mouth of the Niagara until it was 125 feet higher than now. It fairly washed the foot of the cliff at Queenston and Lewiston. Then came a time when the lake fell suddenly through a vertical distance of 250 feet, and its shore retreated to a position now submerged. Numerous minor oscillations were caused by successive shiftings of the point of discharge, and by progressive changes in the attitude of the land, until finally the present outlet was acquired, at which time the Niagara river had its greatest length. It then encroached five miles on the modern domain of Lake Ontario, and began a delta where now the lead-line runs out thirty fathoms.

While the level of discharge was lower than now, the river had different powers as an eroding agent. The rocks underlying the low plain along the margin of the lake are very soft, and where a river flows across yielding rocks, the depth to which it erodes is

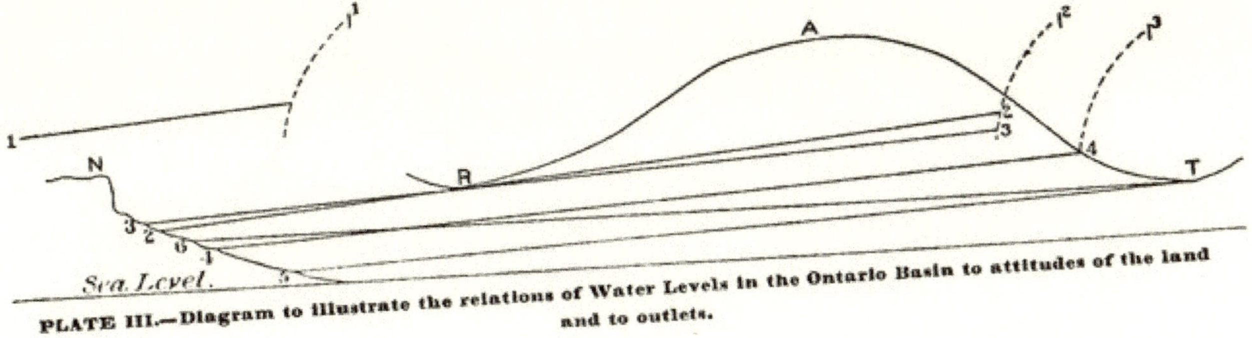

PLATE III.—Diagram to illustrate the relations of Water Levels in the Ontario Basin to attitudes of the land and to outlets.

limited chiefly by the level of its point of discharge. So when the point of discharge of the Niagara river — the surface of the lake to which it flowed — was from 100 to 200 feet lower than now, the river carved a channel far deeper than it could now carve. When afterward the rise of land in the vicinity of the outlet carried the water gradually up to its present position in the basin, this channel was partly filled by sand and other débris brought by the current; but it was not completely filled, and its remarkable present depth is one of the surviving witnesses of the shifting drama of the Ontario. Near Fort Niagara twelve fathoms of water are shown on the charts.

Mr. Warren Upham has made a similar discovery in the basin of the Red River of the North. That basin held a large lake, draining southward to the Mississippi — a lake whose association with the great glacier Upham appropriately signalized by naming it after the apostle of "the glacial theory," Louis Aggassiz. The height of the old Agassiz shore has been carefully measured by Mr. Upham, through long distances, and it is found to rise continuously, though not quite uniformly, toward the north. Similar discoveries have been made in the basins of Erie, Huron and Michigan, and the phenomena all belong approximately to the same epoch. So, while the details remain to be worked out, the general fact is already established that during the epoch of the ice retreat the great plain constituting the Laurentian basin was more inclined to the northward than at present.

It was shown, first in the case of Lake Agassiz, and afterward, as already stated, in the case of Lake Ontario, that the change from the old attitude of the land to the present attitude was in progress during the epoch of the ice retreat. The land was gradually rising to the north or northeast. In each lake basin the water either retreated from its northern margin, so as to lay bare more land, or encroached on its southern margin, or else both these changes occurred together; and in some cases we have reason to believe that the changes were so extensive that the outlets of lakes were shifted from northerly passes to more southerly passes.

To illustrate the effect of the earlier system of land slopes upon the distribution of water in the region of the Great Lakes, I have constructed the map in Pl. IV. It does not postulate the system of levels most divergent from the present system, but a system such as may have existed at the point of time when the last glacial ice was melted from the region. The modern system of drainage is drawn in broken lines; the hypothetic system in full lines, with shading for the lake areas; and a heavier broken line toward the bottom of the map marks the position of the present water-parting at the southern edge of the Laurentian basin.

In the ancient system of drainage, Georgian bay, instead of being a dependency of Lake Huron, is itself the principal lake, and receives the overflow from Huron. It expands toward the northeast so as to include the basin of Lake Nipissing, and its discharge is across a somewhat low pass at the east end of Lake Nipissing, and thence down the Ottawa river to the St. Lawrence. Lake Michigan, instead of communicating with Lake Huron by a strait, forms a tributary lake, discharging its surplus through a river. Lake Superior has the same relations as now, but its overflow traverses a greater distance before reaching Lake Huron. Superior, Michigan, Huron and Georgia constitute a lake system by themselves, independent of Erie and Ontario, and the channel of the Detroit river is dry. Lake Erie and Lake Ontario, both greatly reduced in size, constitute another chain, but their connecting link, the Niagara river, is a comparatively small stream, for the diversion of the upper lakes robs the river of seven-eighths of its tributary area.

Whether this hypothetic state of drainage ever existed, whether the ice retreated from the Nipissing pass while still the changing attitude of the land was such as to turn the Georgian outlet in that direction, are questions not yet answered. But such data as I have at present incline me to the belief that for a time the upper lakes did discharge across the Nipissing pass.

Professor Spencer has described a channel by which Georgian bay once drained across a more southerly pass to the valley of the Trent river, and thence to Lake Ontario.[1] He states that there is an ancient shore-line about Georgian bay associated with this outlet, and that he has traced this line westward and southward until it comes down to the shore of Lake Huron, demonstrating that during the existence of that outlet also, the Detroit river ran

1 Proc. A. A. A. S., 37th Meeting (Cleveland), pp. 198-199.

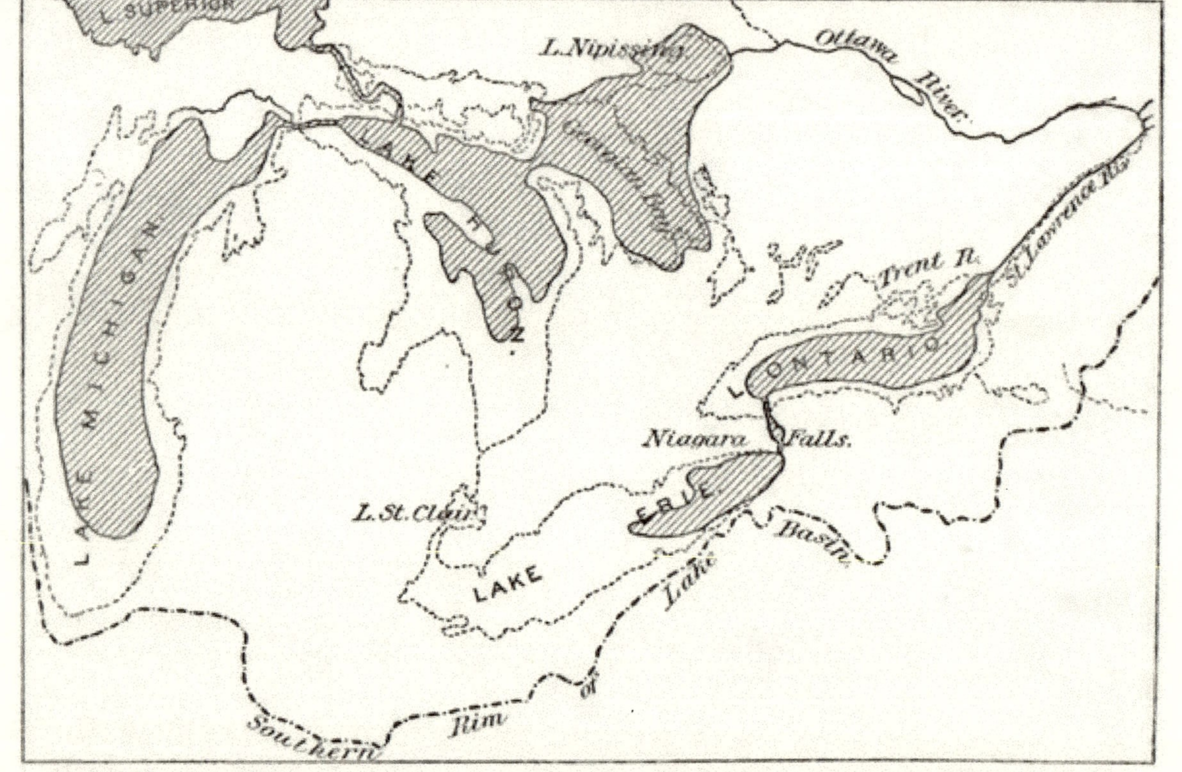

PLATE IV.—Hypothetic hydrography at a date after the melting of the great glacier from the St. Lawrence Valley.

EXPLANATION.—Water-parting in heavy broken line. Modern hydrography in light broken lines. Ancient rivers in full lines. Ancient lakes shaded.

PLATE V.—Hypothetic hydrography at a date before the melting of the great glacier from the St. Lawrence Valley.

EXPLANATION.—Water-parting in heavy broken line. Modern hydrography in light broken lines. Ancient rivers in full lines. Ancient lakes shaded. Ice sheet cross-shaded.

dry. The Trent pass is much higher than the Nipissing pass, so that it appears necessary to assume that during the history of the Trent outlet for the upper lakes, the great glacier still occupied the region of Lake Nipissing, preventing the escape of the water in that direction.

The map in Pl. V represents the system of lakes and outlets at that time. It is largely theoretic, but I believe its general features consistent with our present knowledge of the facts.

Unless I have misunderstood Professor Spencer, Lake Ontario was at high stage in the first part of the epoch of the Trent valley outlet, and was afterward at low stage. I have selected as the date of my map the epoch of the high stage, with the outlet of Ontario at Rome, and have indicated an ice sheet so extensive as to block the way, not only at Lake Nipissing, but at the pass of the Thousand Islands. The date of this map is earlier than the other; it belongs to a time when the northward depression of the land was greater. Lake Erie is represented as less in extent, for its basin in that position would hold less water. Huron and Ontario would likewise be smaller were their waters free to escape over the lowest passes; but the ice blocks the way, and so their waters are raised to the level of higher passes. Of the contemporaneous relations of the upper lakes we know nothing at present. They are drawn as though communicating with Lake Huron, but it is equally possible that they fell into some other drainage system. Here again the Detroit channel was not in use, and the Niagara river was outlet only for the waters of the Erie basin.

Graphic methods are ill adapted to the communication of qualified or indefinite statements. By the aid of a map one can indicate definitely the relation of Albany to other places and things, but he can not say indefinitely that Albany is somewhere in eastern New York, nor can he say, with qualification, that it is probably on the Mohawk river. For this reason I have decided to publish these two maps only after hesitation, because I should greatly regret to produce the impression that the particular configuration of lakes and outlets here delineated has been actually demonstrated. The facts now at command are suggestive rather than conclusive, and when the subject shall have been fully investigated it is to be expected that the maps representing these epochs will exhibit material differences from those I have drawn. The sole point that I wish to develop at this time is the probability that during a portion of the history of the Niagara river, its drainage district — that area from which its water was supplied — was far less than it is at the present time. There is reason to believe that during an epoch which may have been short or long, we can only vaguely conjecture, the Niagara was a comparatively small river.

The characters of the gorge are in general remarkably uniform from end to end. Its width does not vary greatly; its course is flexed but slightly; its walls exhibit the same alternation of soft and hard rocks. But there is one exceptional point. Midway, its course is abruptly bent at right angles. On the outside of the angle there is an enlargement of the gorge, and this enlargement contains a deep pool, called the Whirlpool. At this point, and on this side only, the material of the wall has an exceptional character. At every other point there is an alternation of shales, sandstones and limestones, capped above by an unequal deposit of drift. At this point, limestones, sandstones and shales disappear, and the whole wall is made of drift. Here is a place where the strata that floor the plateau are discontinuous, and must have been discontinuous before the last occupation of the region of the glacier, for the gap is filled by glacial drift.

Another physiographic feature was joined to this by Lyell and Hall. They observed that the cliff limiting the plateau has, in general, a very straight course, with few indentations. But at the town of St. Davids, a few miles west of Queenston, a wide flaring gap occurs. This gap is partly filled by drift, and although the glacial nature of the drift was not then understood, it was clearly perceived by those geologists that the drift-filled break marked the position of a line of erosion established before the period of the drift. Putting together the two anomalies, they said that the drift-filled gap at the Whirlpool belonged to the same line of ancient erosion with the drift-filled gap at St. Davids.[1] Their conclusion has been generally accepted by subsequent investigators, but the interpretation of the phenomena was carried little farther until

1 Travels in North America. By Charles Lyell. New York, 1845. Vol. II, pp. 77-80. Natural History of New York. Geology. Part IV. By James Hall, pp. 389-390.

PLATE VI.—Bird's-eye view of the Niagara Gorge.

the subject was studied by Dr. Julius Pohlman.[1] He pointed out that the upper course of the ancient gorge could not have lain outside the modern gorge. If the course of one gorge lay athwart the course of the other, we should have two breaks in the continuity of the strata, instead of the single one at the Whirlpool. The upper part of the ancient gorge necessarily coincides with a part of the modern gorge; and so, when the cataract, in the progressive excavation of the canyon, reached a point at the Whirlpool where it had no firm rock to erode, it had only to clear out the incoherent earth and boulders of glacial drift. To whatever distance the gorge of the earlier stream extended, the modern river found its laborious task performed in advance.

Let us put together what we have learned of the Niagara history. The river began its existence during the final retreat of the great ice sheet, or, in other words, during the series of events that closed the age of ice in North America. If we consider as a geologic period the entire time that has elapsed since the beginning of the age of ice, then the history of the Niagara river covers only a portion of that period. In the judgment of most students of glacial geology, and, I may add, in my own judgment, it covers only a small portion of that period.

During the course of its history, the length of the river has suffered some variation by reason of the successive fall and rise of the level of Lake Ontario. It was at first a few miles shorter than now; then it became suddenly a few miles longer, and its present length was gradually acquired.

With the change in the position of its mouth there went a change in the height of its mouth; and the rate at which it eroded its channel was affected thereby. The influence on the rate of erosion was felt chiefly along the lower course of the river, between Lewiston and Fort Niagara.

The volume of the river has likewise been inconstant. In early days, when the lakes levied a large tribute on the melting glacier, the Niagara may have been a larger river than now; but there was a time when the discharge from the upper lakes avoided the route by Lake Erie, and then the Niagara was a relatively small stream.

1 Proc. A. A. A. S., 35th Meeting (Buffalo), pp. 221-222.

The great life work of the river has been the digging of the gorge through which it runs from the cataract to Lewiston. The beginning of its life was the beginning of that task. The length of the gorge is in some sense a measure of the river's age. In the main the material dug has been hard limestone and sandstone, interbedded with a coherent though softer shale; but for a part of the distance the material was incoherent drift.

The geologic age of the earth — the time during which its surface has been somewhat as now, divided into land and ocean, subject to endless waste on the land and to endless accumulation of sediment in the ocean, green with verdure and nourishing the varied forms of animal life — this time is of immense duration. Even the units into which geologists divide it, the periods and epochs of their chronology, are themselves of vast duration. Human history is relatively so short, and its units of centuries and years are so exceedingly brief, that the two orders of time are hardly commensurate. Over and over again the attempt has been made to link together the two chronologies, to obtain for the geologic units some satisfactory expression in the units of human history. It can not in fairness be said that all these attempts have failed, for some of them are novel and untested; but, however successful or unsuccessful they may have been, the interest in the subject remains, and no discussion of the history of the Niagara river would be complete without some allusion to its value as a geologic chronometer. It is true we know but little of the ratio the river epoch bears to the extent of the glacial period, or to any longer geologic unit; but yet were we able to determine, even approximately, the time consumed by the river in cutting its gorge, we should render less hazy and vague our conception of the order of magnitude of the units of the earth's geologic history. The problem has been attacked by numerous writers, and the resulting estimates have ranged from three or four thousand years to three or four million years.

The method of reaching a time estimate has been, first, to estimate the present rate of recession — the rate at which the cataract is increasing the length of the gorge; second, to compute, with the aid of this estimate and the known length of the gorge, the time necessary for the entire excavation; and, third, some writers have modified their result by giving consideration to various conditions affecting the rate of erosion during earlier stages of the excavation. The enormous range of the resulting estimates of time has depended chiefly upon the imperfection of data with reference to the present rate of recession of the falls. It is but a few years since measurement of the rate of recession was substituted for bald guessing.

This measurement consists in making surveys and maps of the falls at different times, so that the amount of change in the interval between surveys can be ascertained by comparison of the maps. In 1842 Professor Hall made a survey of the outlines of the falls, and he published, for the use of future investigators, not only the map resulting from the survey, but also the bearings taken with the surveying instrument in determining the principal points of the map.[1] He likewise left upon the ground a number of well-marked monuments to which future surveys could be referred. Thirty-three years later a second survey was made by the United States Army Engineers, and they added still further to the series of bench marks available for future reference. Three years ago. my colleague, Mr. R. S. Woodward, executed a third survey.[2]

Plate VII exhibits the outline of the crest of the falls, together with the brink of the cliff in the vicinity of the falls, as determined by Mr. Woodward in 1886, and also shows a part of the same outline as determined by Professor Hall forty-four years earlier.[3] If both were precise, the area included between the two lines would exactly represent the recession of the Horseshoe and American falls in forty-four years, and the retreat of the cliff face at Goat Island in the same time. I regret to say that there is internal evidence pointing to some defect in one or both surveys, for there are some points at which the Woodward outline projects farther towards the gorge than the Hall outline, and yet we can not believe that any additions have been made to the face of the cliff. Nevertheless, a critical study, not merely of these bare lines on the chart, but also of the fuller data in the surveyors' notes, leads to the belief that the rate of recession in the central part of the Horse-

1 Natural History of New York, Geology, Part IV, pp. 402-403.

2 Science, Vol. VIII, 1886, p. 205.

3 The south side of this chart is placed uppermost (in violation of the conventional rule) so that it may accord with the bird's-eye views.

shoe Fall is approximately determined, and that it is somewhere between four feet and six feet per annum. The amount fallen away at the sides of the Horseshoe is not well determined, but this is of less importance, for such falling away affects the width of the gorge rather than its length, and it is the length with which we are concerned.

The surveys likewise fail to afford any valuable estimate of the rate of retreat of the American Fall, merely telling us that its rate is far less than that of the Horseshoe — a result that might be reached independently by going back in imagination to the time when the two falls were together at the foot of Goat Island, and considering how much greater is the distance through which the Horseshoe Fall has since retreated. The rate of retreat of the central portion of the Horseshoe is the rate at which the gorge grows longer.

Now if we were to divide the entire length of the gorge by the space through which the Horseshoe Fall retreats in a year, we might regard the resulting quotient as expressing the number of years that the falls have been occupied with their work. This is precisely the procedure by which the majority of time estimates have been deduced, but in my jndgment it is not defensible. It implies that the rate of retrogression has been uniform, or, more precisely, that the present rate of retrogression does not differ from the average rate, and this implication is open to serious question. I conceive that future progress in the discussion of the time problem will consist chiefly in determining in what ways the conditions or circumstances that affect the rate of retrogression have varied in past time. In order to discuss intelligently these conditions, it is necessary to understand just what is the process by which the river increases the length of its gorge.

There can be no question that the cataract is the efficient engine, but what kind of an engine is it? What is the principle on which it works?

It has already been stated that the rocks at the falls lie in level layers. The order of succession of the layers has much to do with the nature of the cataract's work. Above all is a loose sheet of drift, but this yields so readily to the wash of the water that we need pay no attention to it at present. Under that is a bed of strong

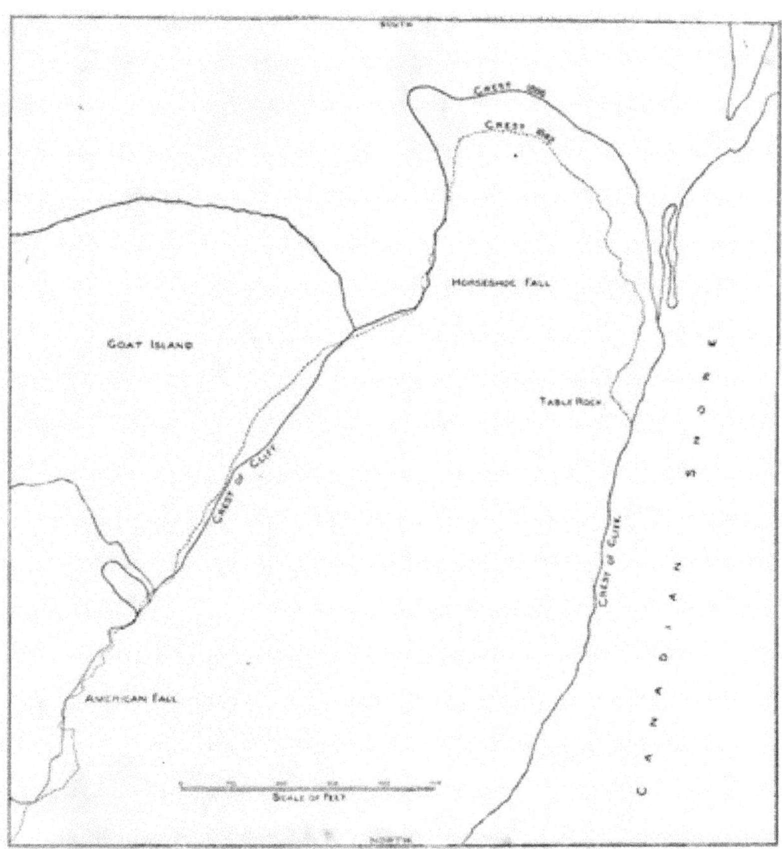

PLATE VII.—Chart of the Cliff Line at the head of the Niagara Gorge, compiled to show the Recession from 1842 to 1886.

EXPLANATION.—Broken line, crest of falls and cliff as mapped by N. Y. State Geol- Survey in 1842. Full line, crest of falls as mapped by the U. S. Geol. Survey in 1886, with other features as mapped by the U S. Lake Survey in 1875.

limestone. This is called the Niagara limestone, and its thickness is eighty feet. Beneath it is a shale, called the Niagara shale, with a thickness of fifty feet; and then for thirty-five feet there is an alternation of limestone, shale and sandstone, known collectively as the Clinton group. This reaches down very nearly to the water's edge. Beneath it, and extending downward for several hundred feet, is a great bed of soft, sandy shale, interrupted, so far as we know, by but a single hard layer, a sandstone ledge, varying in thickness from ten to twenty feet. These are the Medina shales and the Medina sandstone. The profile in Plate VIII indicates that the hard layers project as shelves or steps, and that the softer layers are eaten back. I have been led so to draw them by considerations of analogy only, for underneath the center of the great cataract no observations have been made. We only know that the river leaps from the upper surface of the Niagara limestone and strikes upon the water of the pool. The indicated depth of the pool, too, is a mere surmise, for in that commotion of waters direct observation is out of the question. But where the United States Engineers were able to lower their plummet, a half a mile away, a depth was discovered of nearly 200 feet, and I have assumed that the cataract is scouring as deeply now as it scoured at the time when that part of the gorge was dug.

It is a matter of direct observation that, from time to time, large blocks of the upper limestone fall away into the pool, and there seems no escape from the inference that this occurs because the erosion of the shale beneath deprives the limestone of its support. Just how the shale is eroded, and what is the part played by the harder layers beneath, are questions in regard to which we are much in doubt. In the Cave of the Winds, where one can pass beneath and behind one of the thinner segments of the divided fall, the air is filled with spray and heavier masses of water that perpetually dash against the shale, and though their force in that place does not seem to be violent, it is possible that their continual beating is the action that removes the shaly rock. The shale is of the variety known as calcareous, and as its calcareous element is soluble, it may be that solution plays its part in the work of undermining. What goes on beneath the water of the pool must be essentially different. The Niagara river carries no sediment, and therefore can not scour its channel in the manner of most rivers, but the fragments of the limestone bed that fall into the pool must be moved by the plunging water, else they would accumulate and impede its work; and being moved, we can understand that they become powerful agents of excavation. Water plunging into a pool acquires a gyratory motion, and, carrying detritus about with it, sometimes bores deep holes, even in rocks that are hard. These holes are called technically "pot-holes," and there is much to commend the suggestion that the excavation within the pooi is essentially pot-hole work.[1]

The process which I have described is that which takes place in the central part of the Horseshoe Fall, where the greatest body of water is precipitated. At the margin of the Horseshoe, and also at the American Fall, in which places the body of falling water is much less, the process is different. There is there no pot-hole action and no pool. The fallen blocks of limestone form a low talus at the foot of the cliff, and upon them the force of the descending water is broken and spent. Such of you as have made the excursion through the Cave of the Winds, will recall that though for a few steps you traveled upon an undisturbed rock stratum, one of the layers of the Clinton group, the greater part of the journey lay across large, fallen blocks of limestone, irregularly heaped. Where, then, the volume of falling water is relatively small, the great bed of shale below the Clinton ledges plays no part, and the rate at which the limestone breaks away is determined purely by the rate of erosion of the shale bed lying just beneath it.

The difference between the two processes is of great importance in the present connection, because the two rates of erosion are very different.

I am fully aware that this sketch of the cataract's work is not a satisfactory explanation of the mode of recession, but it yet serves a present purpose, for it renders it possible to point out that the rate of recession is affected by certain factors which may have varied during the earlier history of the river. We see that the process of recession is concerned with a heavy bed of hard rock

1 I am indebted for this suggestion to Mr. W. J. McGee.

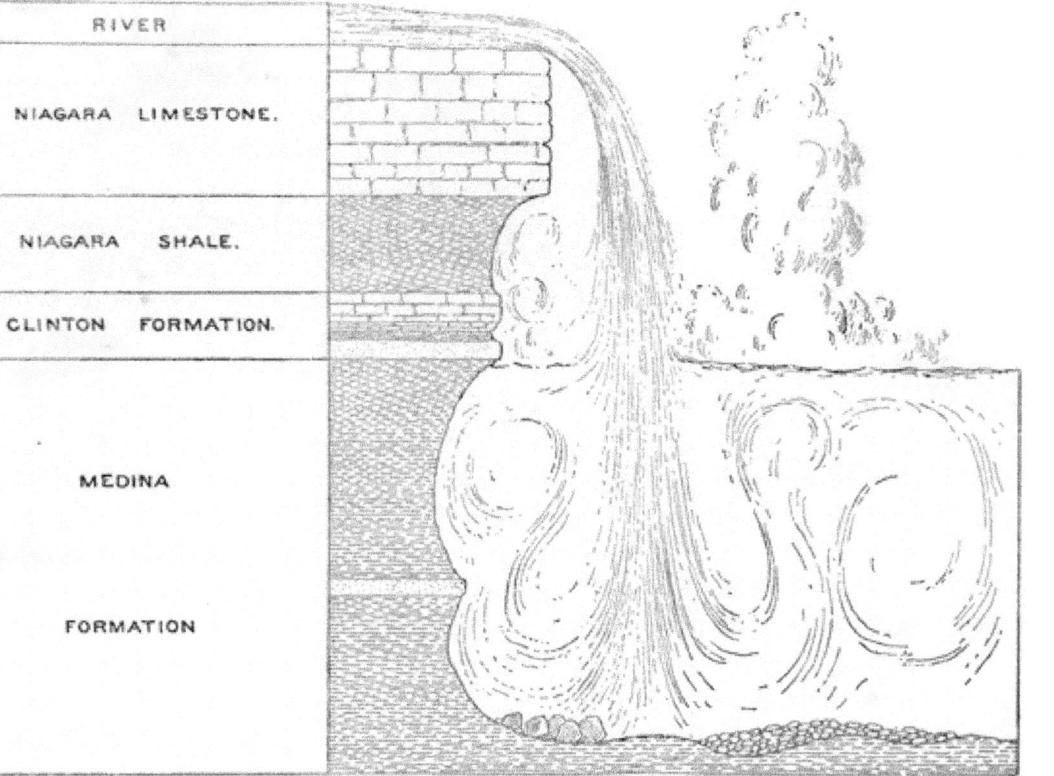

PLATE VIII.—Section of *Niagara Falls*, showing the arrangement of hard and soft strata, and illustrating a theory of the process of erosion.

above, with beds of softer rock beneath, with the force of falling water, and possibly, also, with the solvent power of the water.

Concerning each of these factors a number of pertinent questions may be asked, questions that should certainly be considered, whether they are answered or not, before any solution of the time problem is regarded as satisfactory. To illustrate their pertinence, a few will be propounded.

Question 1. Does the limestone vary in constitution in different parts of the gorge? If its texture or its system of cracks and joints varies, the process of recession may vary in consequence.

Question 2. How does the limestone bed vary in thickness in different parts of the gorge? This question is easily answered, for at all points it is well exposed for measurement.

Question 3. How is the thickness of the limestone related to the rate of recession? This is more difficult. The débris from a very thick bed of limestone would oppose great resistance to the cataract and check its work, The débris from a very thin bed would afford small and inefficient pestles for pot-hole action, and might lead to a slow rate of recession. If the thickness now seen at the cataract were slightly increased or slightly diminished, it is not at once apparent how the rate of recession would be affected, and yet there might be an important difference.

We have seen that the pre-glacial stream whose channel is betrayed at the Whirlpool, removed the Niagara limestone through a portion of the gorge, and

Question 4 asks: Through what portion of the gorge was the Niagara limestone absent when the Niagara river began its work?

Question 5. Does the rock section beneath the limestone — the shale series with its imbedded harder layers — does this vary in different parts of the gorge?

Question 6. Through what distance were the several members of the underlying rock series removed by the action of the pre-glacial stream?

Coming now to consider the force of the falling water, a little consideration serves to show that the force depends on at least three things: The height through which the water falls, the degree of concentration of the stream, and the volume of the river.

The height of the fall is the vertical distance from its crest to the surface of the pool below,

Question 7 asks: How has the height of the crest of the fall varied during the history of recession?

Question 8. How has the height of the base of the fall varied? And this involves a subsidiary question,— to what extent has the excavated gorge, as left by the retreating cataract, been refilled, either by the falling in of fragments from the cliffs, or by contributions of débris brought by the current?

Question 9. What has been the form of the channel at the crest of the fall, from point to point, during the recession? Wherever the channel has been broad, and the water of uniform depth from side to side, the force of the falling water has been applied disadvantageously; wherever the channel has been narrow, or has been much deeper in some parts than in others, the force of the water has been applied advantageously.

There are many ways in which it is possible that the volume of the river was made to differ at early dates from its present volume. During the presence of the ice, there was a different climate, and there were different drainage systems.

Question 10. During the early history of the river, was the annual rainfall on which its water supply depended greater or less than now?

Question 11. Was the evaporation from the basin at that time greater or less than now? It is believed that at the present time the Niagara river receives less than half the water that falls upon its basin in rain and snow, the remainder being returned to the air by evaporation from the lakes, from the surface of the land and from vegetation.

Question 12. Was the water supply increased by ablation? There may have been times when the overlapping edge of the glacier discharged to the Laurentian basin large bodies of water furnished by the melting of ice that had congealed from the clouds of regions far away.

Question 13. Was the drainage area of the river at any time increased through the agency of ice barriers? Just as the Winnipeg basin was made to send its water to the Mississippi, so we can imagine that regions north of the Great Lakes and now tributary to Hudson's bay, had their discharge temporarily turned to Lake Superior and Lake Huron.

On the other hand, we have seen that the discharge of the whole district of the upper lakes was for a time turned away from the Niagara river. Therefore, we ask:

Question 14. To what extent and for what periods was the volume of the river diminished through the diversion of the discharge of the upper lakes?

Assuming all these questions to be answered one by one, and the variations of different sorts determined, it is still necessary to learn the relations of those variations to each other, and so we ask:

Question 15. How have the variations of rock section, the variations of cataract height, the variations of form of channel and the variations of volume been related to one another in point of time? What have been their actual combinations?

Question 16. How have the various temporary combinations of factors affected the process of retreat and the rate of recession.

The tale of questions is not exhausted, but no more are needed if only it has been shown that the subject is not in reality simple, as many have assumed, but highly complex. Some of the questions are, indeed, easily answered. It may be possible to show that others are of small moment. It may even be that careful study of the local features will enable the investigator to infer the process of cataract work at each point from the existing condition of the gorge, and thus relieve him from the necessity of considering such remote questions as the nature of glacial climate and the history of glacial retreat. But after all paring and pruning what remains of the problem will be no bagatelle. It is not to be solved by a few figures on a slate, nor yet by the writing of many essays. It is not to be solved by the cunning discussion of our scant, yet too puzzling knowledge — smoothing away inconvenient doubts with convenient assumptions, and cancelling out, as though compensatory, terms of unknown value that happen to stand on opposite sides of the equation. It is a problem of nature, and like other natural problems demands the patient gathering of many facts, of facts of many kinds, of categories of facts suggested by the tentative theories of to-day, and of new categories of facts to be suggested by new theories.

I have said our problem is but the stepping stone to another problem, the discovery of common units for earth history and human history. The Niagara bridges the chasm in another way, or more strictly, in another sense, for the term of its life belongs to both histories. The river sprang from a great geologic revolution, the banishment of the dynasty of cold, and so its lifetime is a geologic epoch; but from first to last man has been the witness of its toil, and so its history is interwoven with the history of man. The human comrade of the river's youth was not, alas, a reporter with a note-book, else our present labor would be light. He has even told us little of himself. We only know that on a gravelly beach of Lake Iroquois, now the Ridge road, he rudely gathered stones to make a hearth, and built a fire; and the next storm breakers, forcing back the beach, buried and thus preserved, to gratify yet whet our curiosity, hearth, ashes and charred sticks.[1]

In these Darwinian days, we can not deem primeval the man possessed of the Promethean art of fire, and so his presence on the scene adds zest to the pursuit of the Niagara problem. Whatever the antiquity of the great cataract may be found to be, the antiquity of man is greater.

---

1 American Anthropologist, Vol. II, pp. 173-174.

www.ingramcontent.com/pod-product-compliance
Lightning Source LLC
Chambersburg PA
CBHW032141080426
42733CB00008B/1153

* 9 7 8 3 7 4 3 3 3 2 4 2 3 *